The Works of Oliver Heaviside
By Oliver Heaviside

Electromagnetic waves, the propagation of POTENTIAL, and the electromagnetic effects of a moving charge (1888)
by Oliver Heaviside

PART I.

IN connection with the letters of Profs. Poynting and Lodge in *The Electrician*, Nov. 2, 1888, I believe that the following extract from a letter from Sir William Thomson (which I have permission to publish) will be of interest [see Postscript, p. 483, vol. I., to elucidate]:

"I don't agree that velocity of propagation of electric potential is a merely metaphysical question. Consider an electrified globe, A, moved to and fro, with simple harmonic motion, if you please, to fix the ideas. Consider very quickly-acting electroscopes B, B', at different distances from A. If the indications of B, B' were exactly in the same phase, however their places are changed, the velocity of propagation of electric potential would be infinite; but if they showed differences of phase, they would demonstrate a velocity of propagation of electric potential.

"Neither is velocity of propagation of 'vector-potential' metaphysical. It is simply the velocity of propagation of electromagnetic force the velocity of electromagnetic waves', in fact."

Taking the second point first, it is, I think, clear that if by the propagation of vector-potential is to be *understood* that of electric and magnetic disturbances, it is merely the mode of expression that is in question. I am myself accustomed to mentally picture the electric and magnetic forces or fluxes, and their propagation, which takes place at the speed of light or thereabouts, because they give the most direct representation of the state of the medium, which, I think, must be agreed is the real physical subject of propagation. But if we regard the vector-potential directly, then we can only get at the state of the medium by complex operations, and we really require to know the vector-potential both as a function of position and of time, for its space-variation has to furnish the magnetic force, and its time-variation the electric force; besides which, there is sometimes the space-variation of a scalar potential in addition to be regarded, before we can tell what the electric force is. Besides this roundaboutness, it implies a knowledge of the full solution, and if we do not possess it, it is much simpler to think of the propagation of the electric and magnetic disturbances, and I find that this method works out much more easily in the solution of problems.

The other question will, I believe, be found to be ultimately of precisely the same nature. Start with the sphere A at rest, and the field steady, and consider two external points, P and P', at different distances. The electric force at them has different values, and the whole field has a potential. But now give the sphere a displacement, and bring it to rest again in a new position. Is the readjustment of potential instantaneous? I should say, Certainly not, and describe what happens thus. When the sphere is moved, magnetic force is generated at its boundary (lines circles of latitude, if the axis be the line of motion), and with it there is necessarily disturbance of electric force. The two together make an electromagnetic wave, which goes out from the sphere at the speed of light, and at the front of the wave we have $E=\mu v H$, where E is the electric and H the magnetic force intensity. Before the front reaches P or P' we have the electric field represented by the potential function, but after that it cannot be so represented until the magnetic force has wholly disappeared, when again we have a steady field representable by a potential function. It is difficult to see how to plainly differentiate any propagation of potential *per se*.

If the motion is simple-harmonic, there is a train of outward waves and no potential. I imagine that an electroscope, if infinitely sensitive and without reactions, would register the actual state of the electric field, irrespective of its steadiness. By an electroscope, as this is a purely theoretical question, I understand the very simplest one, a very small charge at a point; or, say, the unit charge, the force on which is the electric force of the field.

When these things are closely examined into, if the facts as regards the propagation of disturbances (electric and magnetic) are agreed on, the only subject of question is the best mode of expressing them, which I believe to be in terms of the forces, not potentials.

But there really is infinite speed of propagation of potential sometimes; on examination, however, it is found to be nothing more than a mathematical fiction, nothing else being propagated at the infinite speed.

It will be understood that I preach the gospel according to my interpretation of Maxwell, and that any modification his theory of the dielectric may receive may involve a fresh kind of propagation at present not in question.

Nov. 5, 1888.

PART II.

The question raised by Prof. S. P. Thompson (in *The Electrician*, Nov. 16, 1888, p. 54) as to whether the motion of an uncharged dielectric through a field of electric force produces magnetic effects must, I think, be undoubtedly answered in the affirmative. As the distribution of displacement varies, its time-variation is the electric current, with determinable magnetic force to match. When the speed of motion is a small fraction of that of light, we may regard the displacement as having at every moment its proper steady distribution, so that there is no difficulty in estimating the magnetic effects, except, it may be, of a merely mathematical character. For instance, the case of a sphere moving in a field which would be uniform were the sphere absent, may be readily attacked, and does perfectly well to illustrate the general nature of the action.

But if the moved dielectric have the same electric permittivity as the surrounding medium, so that there is no difference made in the steady distribution, the question which may be now raised as to the possible production of transient disturbances is one to which the above theory does not present any immediate answer. I believe that the body will be magnetized transversely to the electric displacement and the velocity. [The motional magnetic force is referred to.]

Another question, somewhat connected, is contained in Prof. Poynting's suggestion (in letter to Prof. Lodge, *The Electrician*, p. 829, vol. XXI.) that electric displacement may possibly be produced without magnetic force by the agency of pyroelectricity. But, whatever the agency, it would, I conceive, be a new fact—quite outside Maxwell's theory legitimately developed. We may have subsidence of electric displacement without magnetic force; but I cannot see any way to produce it.

But the main subject of this communication is the electromagnetic effect of a moving charge. That a moving charge is equivalent to an electric current-element is undoubted, and to call it a convection-current, as Prof. S. P. Thompson does, seems reasonable. The true current has three components, thus,

$$\mathrm{curl}\ \mathbf{H} = 4\pi(\mathbf{C} + \dot{\mathbf{D}} + \rho\mathbf{u}),$$

where $\mathbf{H}$ is the magnetic force, $\mathbf{C}$ the conduction-current, $\mathbf{D}$ the displacement, and ρ the volume-density of electrification moving with velocity $\mathbf{u}$. The addition of the term $\rho\mathbf{u}$ is, I presume, the extension made by Prof. Fitzgerald to which Prof. S. P. Thompson refers. At any rate, I can at present see no other.

There are several ways of arriving at the conclusion that a moving charge must be regarded as an electric current; but, when that is admitted, we are very far from knowing what its magnetic effect is. No cut-and-dried statement of it can be made, because it varies according to circumstances. The magnetic field, whatever it be in a given case, is not that of a current-element (supposing the charge to be at a point), for that is anti-Maxwellian, but is that of the actual system of electric current, which is variable.

Thus, in the case of motion at a speed which is a small fraction of that of light, the magnetic field (as found by Prof. J. J. Thomson) is the same as that of Ampère's current-element represented by $\rho\mathbf{u}$; that is, a current-element whose direction is that of u and whose moment is ρu, if u is the tensor of **u** (understanding by "moment," current-density × volume); but the true current to correspond bears the same relation to the current-element as the induction of an elementary magnet bears to its magnetic moment. The magnetic energy due to the motion of a charge q upon a sphere of radius a in a medium of inductivity μ, at a speed u which is only a very small fraction of that of light, is expressed by $\frac{1}{3}\mu q^2 u^2/a$. But if the speed be not a small fraction of that of light, the result is very different. Increasing the speed of the charge causes not merely greater magnetic force but changes its distribution altogether, and with it that of the electric field. It is no use discussing the potential. There is not one. The magnetic field tends to concentrate itself towards the equatorial plane, or plane through the charge perpendicular to the line of motion. When the speed equals that of light itself this process is complete, and the is simply a plane wave (electromagnetic).

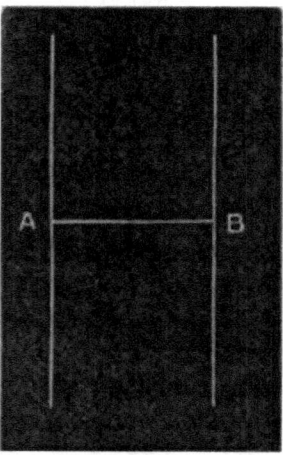

Since a charge at a point gives infinite values, it is more convenient to distribute it. Let it be, first, of linear density q along a straight line AB, moving in its own line at the speed of light. Then the field is contained between the parallel planes through A and B perpendicular to AB, and is completely given by

$$E/\mu v = H = 2qu/r,$$

where E and H are the intensities of the electric and magnetic forces at distance r from AB. The lines of **E** radiate uniformly from AB in all directions parallel to the planes; those of H are everywhere perpendicular to those of **E**, or are circles centred upon AB. Outside this electromagnetic wave there is no disturbance. I should remark that the above is a description of the *exact* solution. It is, of course, nothing like the supposed field of a current-element AB.

To still further realize, we may substitute a cylindrical distribution for the linear, and then, again, terminate the lines of **E** on another cylindrical surface between the bounding planes. To find the resulting distributions of E and H (always perpendicular) may be done by super-imposition of the elementary solutions, or by solving a bidimensional problem in a well-known manner.

Those who are acquainted with my papers in this journal will recognise that what we have arrived at is simply the elementary plane wave travelling along a distortionless circuit. All roads lead to Rome!

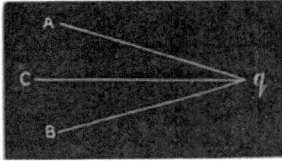

Returning to the case of a charge q at a point moving through a dielectric, if the speed of motion exceeds that of light, the disturbances are wholly left behind the charge, and are confined within a cone, A q B. The charge is at the apex, moving from left to right along C q. The semi-angle, θ, of the cone, or the angle A q C, is given by

$$\sin\ \theta = v/u,$$

where v is the speed of light, and u that of the charge. The magnetic lines are circles round the axis, or line of motion. The displacement is away from q, of course, and of total amount q, but not uniformly distributed within the cone. The electric current is towards q in the inner part of the cone, and away from q in the outer.

It will be seen that the electric stress tends to pull the charge back. Therefore, applied force on q in direction C q is required to keep up the motion. Its activity is accounted for by the continuous addition at a uniform rate which is being made to the electric and magnetic energies at q. For the motion at the wave-front, at any point on A q or B q is perpendicularly outward, not towards q. Whilst the cone is thus expanding all over, the forward motion of q continually renews the apex, and keeps the shape unchanged.

Steady motion alone is assumed.

To avoid misconception I should remark that this is not in any way an account of what would happen if a charge were impelled to move through the ether at a speed several times that of light, about which I know nothing; but an account of what would happen if Maxwell's theory of the dielectric kept true under the circumstances, and if I have not misinterpreted it. [See footnote on p. 516, later.]

Nov. 18, 1888.

PART III.

All disturbances being propagated through the dielectric ether at the speed of light, when, therefore, a charge is in motion through the medium, the discussion of the effects produced naturally involves the consideration of three cases, those in which the speed u of the charge is less than, or equal to, or greater than v, that of light.

In a previous communication [Part II. above], I gave the complete and very simple solution of the intermediate case of equality of speeds. A formal demonstration is unnecessary, as the satisfaction of the necessary conditions may be immediately tested.

But I was not then aware that the case $u<v$ admitted of being presented in a nearly equally-simple form. That such is the fact is rather surprising, for it is very exceptional to arrive at simple results, and these now in question are sufficiently free from complexity to take a place in text-books of electricity.

Let the axis of z be the line of motion of the charge q at speed u. Everything is symmetrical with respect to this axis. The lines of electric force are radial out from the charge. Those of magnetic force are circles about the axis. The two forces are perpendicular. Having thus settled the directions, it only remains to specify their intensities at any point P distant r from the charge, the line r making an angle θ with the axis. Let E be the intensity of the electric, and H of the magnetic force. Then, if c is the permittivity and μ the inductivity, such that $\mu c v^2 = 1$, we have

$$(u<v)\begin{cases} cE = \dfrac{\frac{q}{r^2}\left(1-\frac{u^2}{v^2}\right)}{\left(1-\frac{u^2}{v^2}\sin^2\theta\right)^{\frac{3}{2}}}, & \cdots\cdots\cdots \quad (A) \\[2em] H = cEu\ \sin\theta. & \cdots\cdots\cdots\cdots\cdots \quad (B) \end{cases}$$

That (A), (B) represent the complete solution may be proved by subjecting them to the proper tests. Premising that the whole system is in steady motion at speed u, we have to satisfy the two fundamental laws of electromagnetism:—

(1). (Faraday's law). The electromotive force of the field [or voltage] in any circuit equals the rate of decrease of the induction through the circuit (or the magnetic current × -4π).

(2). (Maxwell's law). The magnetomotive force of the field [or gaussage] in any circuit equals the electric current × 4π through the circuit.

Besides these, there is continuity of the displacement to be attended to. Thus:—

(3). (Maxwell). The displacement outward through any surface equals the enclosed charge.

Since (A) and (B) satisfy these tests, they are correct. And since no unrealities are involved, there is no room for misinterpretation.

When u/v is very small, we have, approximately,

$$cE=\frac{q}{r^2}, \qquad H=\frac{qu}{r^2}\sin\theta$$

representing Prof. J. J. Thomson's solution—that is, the lines of displacement radiate uniformly from the charge, and the magnetic force is that of the corresponding displacement-currents together with the moving charge regarded as a current-element of moment qu. Instantaneous action through the medium is involved—that is, to make the solution quite correct.

That the lines of electric force should remain straight as the speed of the charge is increased is itself a rather remarkable result. Examining (A), we see that the effect of increasing u is to concentrate the displacement about the equatorial plane $\theta=\frac{1}{2}\pi$.Self-induction does it. In the limit, when $u=v$, the numerator vanishes, making $E=0$, $H=0$ everywhere except at the plane mentioned, where, by reason of the denominator becoming infinitely small in comparison with the numerator, the displacement is all concentrated in a sheet, and with it the induction, forming a plane electromagnetic wave, as described (and realized) in my previous communication.

If we terminate the field described in (A) and (B) on a spherical surface of radius a, instead of continuing it up to the charge q at the origin, we have the case of a perfectly conducting sphere of radius a possessing a total charge q, moving steadily at speed u through the dielectric ether. As the speed is increased to v, the charge all accumulates at the equator of the sphere. [See footnote on p. 514, later.]

But after that? This brings us to the third case of $u>v$, and here I have so-far failed to find any solution which will satisfy all the necessary conditions without unreality. The description at the close of Part II. must therefore be received as a suggestion, at present unconfirmed I hope to consider the matter in a future communication.

P.S.—In a recent number Mr. W. P. Granville raised the question of action through a medium being only action at a short distance instead of a long one, and asked for instruction. His inquiry has elicited no response. This is not, however, because there is nothing to be said about it. The matter did not escape the notice of the "ant-distance-action sage." My own opinion is that the question involved is, if not metaphysical, dangerously near to being so; consequently, whole books might be devoted to it. At present, however, I think it is more useful to try to find out what happens, and to construct a medium to make it happen; after that, perhaps, the matter referred to may be more advantageously discussed. The well of truth is bottomless.

PART IV.

In previous communications [above] I have discussed this matter. Referring to the case of steady rectilinear motion, I gave a description of the result when the speed of the charge exceeds that of light, obtained mainly by general reasoning, and stated my inability to find a solution to represent it. The displacement cannot be outside a certain cone of semi-vertical angle whose sine equals the ratio v/u of the speed of light to that of the charge, which is at the apex.

In the *Phil. Mag.* for July, 1889, Prof. J. J. Thomson has examined this question. Like myself, he fails to find a solution within the cone; but concludes that the displacement is confined to its surface. If so, it must form, along with the magnetic induction, an electromagnetic wave. But it may be readily seen that such a wave is impossible, having no stability.

For as the charge moves from A to B, a given surface-element, C, would move to D. In doing so its area would vary directly as its distance from the apex, and the energy in the element would therefore vary inversely as its distance from the apex, and the forces, electric and magnetic, would therefore vary inversely as the square root of the distance from the apex, instead of inversely as the distance,

which is obviously necessary in order that the displacement may be confined to the surface. This conflict of conditions constitutes instability. In the *Phil. Mag.* for April, 1889, I suggested that whilst there must be a solution of some kind, one representing a *steady* state was impossible. This conclusion is confirmed by the failure of Prof. Thomson's proposed surface-wave to keep itself going.

Prof. Thomson, who otherwise confirms my results, has also extended the matter by supposing that the medium itself is set in motion, as well as the electrification. This is somewhat beyond me. I do not yet know certainly that the ether can move, or its laws of motion if it can. Fresnel thought the earth could move through the ether without disturbing it; Stokes, that it carried the ether along with it, by giving irrotational motion to it. Perhaps the truth is between the two. Then there is the possibility of holes in the ether, as suggested by a German philosopher. When we get into one of these holes, we go out of existence. It is a splendid idea, but experimental evidence is much wanting.

But if we consider that the medium supporting the electric and magnetic fluxes is really set moving when a body moves, and assume a particular kind of motion, it is certainly an interesting scientific question to ask what influence the motion exerts on the electromagnetic phenomena. I do not, however, think that any new principles are involved.

The general connections of **E** and **H**, referred to fixed space without conductivity, being

$$\mathrm{curl}(\mathbf{e}-\mathbf{E})=\mu p \mathbf{H}, \quad \text{-----------------} \quad (1)$$

$$\mathrm{curl}(\mathbf{H}-\mathbf{h})=cp\mathbf{E}, \quad \text{-----------------} \quad (2)$$

where p stands for d/dt and $\mathbf{e}$ and $\mathbf{h}$ are the impressed parts of $\mathbf{E}$ and $\mathbf{H}$; if there is also motion of electrification, we have to consider it to constitute a convection-current a part of the true current, and so make (2) become

$$\mathrm{curl}(\mathbf{H}-\mathbf{h})=cp\mathbf{E}+4\pi\rho\mathbf{u}, \quad \text{-----------------} \quad (3)$$

where ρ is the density of electrification, whose velocity is $\mathbf{u}$. [See Part II.] It now remains to specify $\mathbf{e}$ and $\mathbf{h}$. They are zero when the medium supporting the fluxes is at rest. But if it moves, and its velocity is $\mathbf{w}$, there is, first, the electric force due to motion in a magnetic field,

$$\mathbf{e}=\mu V\mathbf{w}\mathbf{H}, \quad \text{-----------------} \quad (4)$$

which is well known; and next the magnetic force due to motion in an electric field,

$$\mathbf{h}=cV\mathbf{E}\mathbf{w}, \quad \text{-----------------} \quad (5)$$

which is not so well known. (First, I believe, given by me in the third Section of "Electromagnetic Induction and its Propagation," *The Electrician*, January 24, 1885 [vol. I., p. 446], again, obtained in a different way in Section XXIL, January 15, 1886 [vol. I., p. 546]; see also *Phil. Mag.*, August, 1886 [vol. II., Art. L.], and an example of the use of (4) and (5) in *The Electrician*, April 12, 1889, p. 683 [vol. II., Art. LI.].)

The mechanical force called by Maxwell the "electromagnetic force" is $V\mathbf{CB}$, where $\mathbf{C}$ is the true current and $\mathbf{B}$ the induction. It is the force on the matter supporting electric current. Let it move. If $\mathbf{w}$ is its velocity, the activity of the force is

$$\mathbf{w}V\mathbf{CB}=\mathbf{C}V\mathbf{Bw}=-\mathbf{eC}. \quad \text{-----------------} \quad (6)$$

Similarly, as I obtained in Section XXII. above referred to, there is a mechanical force (the magneto-electric) on matter supporting magnetic current $\mathbf{G}=\mu p\,\mathbf{H}/4\pi$, expressed by $4\pi V\mathbf{DG}$, and its activity is

$$4\pi\mathbf{w}V\mathbf{DG}=4\pi\mathbf{G}V\mathbf{wD}=-\mathbf{hG}. \quad \text{-----------------} \quad (7)$$

Of course $\mathbf{e}$ and $\mathbf{h}$ are reckoned as impressed forces, which is the reason of the change of sign. *Their activities are* $\mathbf{eC}$ *and* $\mathbf{hG}$.

It should be remarked further, that the above expressions for $\mathbf{e}$ and $\mathbf{h}$ are not *certain*. For I have shown that the sources of all disturbances are the lines of curl of the impressed forces (*Phil. Mag.*, Dec., 1887) [vol. II., p. 362], and that the fluxes produced depend solely upon the curls of $\mathbf{e}$ and $\mathbf{h}$, both as regards the steady fluxes and the variable ones leading to them. We may, therefore, use any other expressions for $\mathbf{e}$ and $\mathbf{h}$ which have the same curls as the above. And, in fact, we see that equations (1) and (2) only contain their curls.

Equations (1) and (3), with $\mathbf{e}$ and $\mathbf{h}$ defined by (4) and (5), therefore enable us to determine the effect of the moving medium. Prof. Thomson also arrives at (4) and (5), and at the "magneto-electric force," in his paper to which I have referred, by an entirely different method. And to show how well things fit together, he concludes, from the consideration of the moving medium, that a moving electrified surface is a current-sheet, which is another way of saying that a convection current is a part of the true current, as expressed in (3). I must, however, disagree with Prof. Thomson's assumption that the motion must be irrotational. It would appear, by the above, that this limitation is unnecessary.

As an example, and to introduce a new point, take the case of a charge q moving at speed u along the axis of z. It will come to the same thing if we keep the charge at rest, and move the medium the other way. We then use the equations (1) and (2), and in them use (4) and (5) with $\mathbf{w}=-\mathbf{u}$. Now when the steady state is arrived at, we have $p=0$, so (1) and (2) become

$$\text{curl}(\mu\mathbf{V}\mathbf{H}\mathbf{u}-\mathbf{E})=0, \quad \text{...(8)}$$

$$\text{curl}(\mathbf{H}-c\mathbf{V}\mathbf{u}\mathbf{E})=0. \quad \text{...(9)}$$

In addition, the divergence of $\mathbf{D}$ must be q at the origin, and the divergence of $\mathbf{B}$ must be zero. The latter gives, applied to (9),

$$\mathbf{H}=c\mathbf{V}\mathbf{u}\mathbf{E}, \quad \text{...(10)}$$

which gives $\mathbf{H}$ fully in terms of $\mathbf{E}$. Eliminate $\mathbf{H}$ from (8) by means of (10), and we get

$$\text{curl}(\mu c\mathbf{V}\mathbf{u}\mathbf{V}\mathbf{E}\mathbf{u}-\mathbf{E})=0, \quad \text{...(11)}$$

or

$$\text{curl}\left[\frac{u^2}{v^2}(\mathbf{E}-E_3\mathbf{k})-\mathbf{E}\right]=0, \quad \text{...(12)}$$

where E_3 is the z-component of $\mathbf{E}$ and $\mathbf{k}$ a unit vector along $\mathbf{z}$; or, integrating, and writing the three components,

$$E_1=-\frac{dP}{dx}, \quad E_2=-\frac{dP}{dy}, \quad E_3=-\left(1-\frac{u^2}{v^2}\right)\frac{dP}{dz}, \quad \text{..(13)}$$

where P is a scalar potential. Here is the new point. There is a potential, of a peculiar kind. The displacement due to the moving charge is distributed in precisely the same way as if it were at rest in an eolotropic medium, whose permittivity is c in all directions transverse to the line of motion, but is smaller, viz., $c(1-u^2/v^2)$, along that line and parallel to it. The potential P is given by

$$P=\frac{q}{c\left[(x^2+y^2)(1-u^2/v^2)+z^2\right]^{\frac{1}{2}}}. \quad \text{.............................(14)}$$

It is a particular case of eolotropy. In general, c_1, c_2, c_3, the principal permittivities, are all unequal. Then, with q at the origin, the potential is

$$P=\frac{q}{(c_1c_2c_3)^{\frac{1}{2}}\left(\frac{x^2}{c_1}+\frac{y^2}{c_2}+\frac{z^2}{c_3}\right)^{\frac{1}{2}}}. \quad \text{.............................(15)}$$

Observe that although the electric force in the substituted problem of a charge at rest in an eolotropic medium is the slope of a potential; yet it is not so when the medium is isotropic, and moves past the fixed charge, or *vice versâ*, although the distributions of displacement are the same.

When $u=v$, we abolish the permittivity along the z-axis in the substituted case, so that the displacement must be wholly transverse. We then have the plane electromagnetic wave. When u is greater than v it makes the permittivity negative along z; this is an impossible electrical problem, and furnishes another reason for supposing that there can be no steady state in the corresponding electromagnetic problem.

It now remains to find what *would* happen if electrification were conveyed through a medium faster than the natural speed of propagation of disturbances. There is the cone; but what takes place within it?

Aug. 25, 1889.

On the Electromagnetic Effects due to the Motion of ELECTRIFICATION through a Dielectric (1889)

by Oliver Heaviside

Theory of the Slow Motion of a Charge.

1. THE following paper consists of, first, a short discussion of the theory of the *slow* motion of an electric charge through a dielectric, having for object the possible correction of previously published results. Secondly, a discussion of the theory of the electromagnetic effects due to motion of a charge at any speed, with the development of the complete solution in finite form when the motion is steady and rectilinear. Thirdly, a few simple illustrations of the last when the charge is distributed.

Given a steady electric field in a dielectric, due to electrification. It is sufficient to consider a charge q at a point, as we may readily extend results later. If this charge be shifted from one position to another, the displacement varies. In accordance, therefore, with Maxwell's inimitable theory of a dielectric, there is electric current produced. Its time-integral, which is the total change in the displacement, admits of no question; but it is by no means an elementary matter to settle its rate of change in general, or the electric current. But should the speed of the moving charge be only a very small fraction of that of the propagation of disturbances, or that of light, it is clear that the accommodation of the displacement to the new positions which are assumed by the charge during its motion is practically instantaneous in its neighbourhood, so that we may imagine the charge to carry about its stationary field of force rigidly attached to it. This fixation of the displacement at any moment definitely fixes the displacement-current. We at once find, however, that to close the current requires us to regard the moving charge itself as a current-element, of moment equal to the charge multiplied by its velocity; understanding by moment, in the case of a distributed current, the product of current-density and volume. The necessity of regarding the moving charge as an element of the "true current" may be also concluded by simply considering that when a charge q is conveyed *into* any region, an equal displacement simultaneously leaves it through its boundary.

Knowing the electric current, the magnetic force to correspond becomes definitely known if the distribution of inductivity be given; and when this is constant everywhere, as we shall suppose now and later, the magnetic force is simply the circuital vector whose curl is 4π times the electric current; or the vector-potential of the curl of the current; or the curl of the vector-potential of the current, etc., etc. Thus, as found by J. J. Thomson,[1] the magnetic field of a charge moving at a speed which is a small fraction of that of light is that which is commonly ascribed to a current-element itself. I think it, however, preferable to regard the magnetic field as the primary object of attention; or else to regard the complete system of closed current derived from it by taking its curl as the unit, forming what we may term a rational current-element, inasmuch as it is not a mere mathematical abstraction, but is a complete dynamical system involving definite forces and energy.

2. Let the axis of z be the line of motion of the charge q at the speed u; then the lines of magnetic force $\mathbf{H}$ are circles centred upon the axis, in planes perpendicular to it, and its tensor H at distance r from the charge, the line $\mathbf{r}$ making an angle θ with the axis, is given by

$$H = \frac{q}{r^2} u \sin \theta = cEuv, \quad \text{..(1)}$$

where $v = \sin \theta$, E the intensify of the radial electric force, c the permittivity such that $\mu_0 c v^2 = 1$, if μ_0 is the other specific quality of the medium, its inductivity, and v is the speed of propagation.

Since, under the circumstance supposed of u/v being very small, the alteration in the electric field is insensible, and the lines of **E** are radial, we may terminate the fields represented by (1) at any distance $r = a$ from the origin. We then obtain the solution in the case of a charge q upon the surface of a conducting sphere of radius a, moving at speed u. This realization of the problem makes the electric and magnetic energies finite. Whilst, however, agreeing with J. J. Thomson in the fundamentals, I have been unable to corroborate some of his details; and since some of his results have been recently repeated by him in another place,[2] it may be desirable to state the changes I propose, before proceeding to the case of a charge moving at any speed.

The Energy and Forces in the Case of Slow Motion.

3. First, as regards the magnetic energy, say T. This is the space-summation $\Sigma \mu_0 H^2/8\pi$; or, by (1),[3],

$$T = \frac{\mu_0 q^2 u^2}{8\pi} \int \int \int \frac{v^2}{r^2} dr \ d\mu \ d\phi = \frac{\mu_0 q^2 u^2}{3a}. \qquad \cdots\cdots\cdots\cdots (2)$$

The limits are such as include all space outside the sphere $r=a$. The coefficient $\frac{1}{3}$ replaces $\frac{2}{15}$.

4. Next, as regards the mutual magnetic energy M of the moving charge and any external magnetic field. This is the space-summation $\Sigma \mu_0 H_0 H/4\pi$, if H_0 is the external field; and, by a well-known transformation it is equivalent to $\Sigma A_0 \Gamma$, if A_0 is any vector whose curl is $\mu_0 H_0$, whilst Γ is the current-density of the moving system. Further, if we choose A_0 to be circuital, the polar part of Γ will contribute nothing to the summation, so that we are reduced to the volume-integral of the scalar product of the circuital A_0 of the one system and the density of the convection-current in the other. Or, in the present case, with a single moving charge at a point, we have simply the scalar product $A_0 u \ q$ to represent the mutual magnetic energy; or

$$M = A_0 u q, \qquad \cdots\cdots\cdots\cdots\cdots\cdots\cdots (3)$$

which is double J. J. Thomson's result.

5. When, therefore, we derive from (3) the mechanical force on the moving charge due to the external magnetic field, we obtain simply Maxwell's "electromagnetic force" on a current-element, the vector product of the moment of the current and the induction of the external field; or, if F is this mechanical force,

$$F = \mu_0 q V u H_0, \qquad \cdots\cdots\cdots\cdots (4)$$

which is also double J. J. Thomson's result. Notice that in the application of the "electromagnetic force" formula, it is the moment of the convection-current that occurs. This is not the same as the moment of the true current, which varies according to circumstances; for instance, in the case of a small dielectric sphere uniformly electrified throughout its volume, the moment of the true current would be only $\frac{2}{3}$ of that of the convection-current.

The application of Lagrange's equation of motion to (3) also gives the force on q due to the electric field so far as it can depend on M; that is, a force

$$-q A_0,$$

where the time-variation due to all causes must be reckoned, except that due to the motion of q itself, which is allowed for in (4). And besides this, there may be electric force not derivable from A_0, viz.

$$-q \nabla \Psi_0,$$

where Ψ_0 is the scalar potential companion to $\mathbf{A}_0$.

6. Now if the external field be that of another moving charge, we shall obtain the mutual magnetic energy from (3) by letting $\mathbf{A}_0$ be the vector-potential of the current in the second moving system, constructed so as to be circuital. Now the vector-potential of the convection-current $q\,\mathbf{u}$. is simply $q\,\mathbf{u}/r$; this is sufficient to obtain the magnetic force by curling; but if used to calculate the mutual energy, the space-summation would have to include every element of current in the other system. To make the vector-potential circuital, and so be able to abolish this work, we must add on to $q\,\mathbf{u}/r$ the vector-potential of the displacement current to correspond. Now the complete current may be considered to consist of a linear element $q\,\mathbf{u}$ having two poles; a radial current outward from the $+$ pole in which the current-density is $qu/4\pi r_1^2$; and a radial current inward to the $-$ pole, in which the current-density is $-qu/4\pi r_2^2$; where r_1 and r_2 are the distances of any point from the poles. The vector-potentials of these currents are also radial, and their tensors are $\frac{1}{2}qu$ and $-\frac{1}{2}qu$. We have now merely to find their resultant when the linear element is indefinitely shortened, add on to the former $q\,\mathbf{u}/r$, and multiply by μ_0, to obtain the complete circuital vector-potential of $q\,\mathbf{u}$, viz.:—

$$\mathbf{A}=\mu_0\,q\Big(\tfrac{u}{r}-\tfrac{1}{2}u\nabla\tfrac{dr}{ds}\Big), \quad\text{.............................}(5)$$

where r is the distance from q to the point P when $\mathbf{A}$ is reckoned, and the differentiation is to s, the axis of the convection-current. Both it and the space-variation are taken at P. The tensor of $\mathbf{u}$ is u. Though different and simpler in form (apart from the use of vectors) this vector-potential is, I believe, really the same as the one used by J. J. Thomson. From it we at once find, by the method described in § 4, the mutual energy of a pair of point-charges q_1 and q_2, moving at velocities $\mathbf{u}_1$ and $\mathbf{u}_2$, to be

$$M=\mu_0 q_1 q_2\Big(\tfrac{u_1 u_2}{r}-\tfrac{1}{2}u_1 u_2\tfrac{d^2 r}{ds_1 ds_2}\Big), \quad\text{.......................}(6)$$

when at distance r apart. Both axial differentiations are to be effected at one end of the line r.

As an alternative form, let ε be the angle between $\mathbf{u}_1$ and $\mathbf{u}_2$, and let the differentiation to s_1 be at ds_1 that to s_2 at ds_2, as in the German investigations relating to current-elements; then[4]

$$M=\mu_0 q_1 q_2\, u_1 u_2\Big(\tfrac{\cos\epsilon}{r}+\tfrac{1}{2}\tfrac{d^2 r}{ds_1 ds_2}\Big). \quad\text{.......................}(7)$$

Another form, to render its meaning plainer. Let λ_1, μ_1, ν_1 and λ_2, μ_2, ν_2 be the direction-cosines of the elements referred to rectangular axes, with the x-axis, to which λ_1 and λ_2 refer, chosen as the line joining the elements. Then[5]

$$M=\tfrac{\mu_0 q_1 q_2 u_1 u_2}{2r}(2\lambda_1\lambda_2+\mu_1\mu_2+\nu_1\nu_2). \quad\text{..................}(8)$$

J. J. Thomson's estimate is[6]

$$M=\tfrac{1}{3}\mu_0 q_1 q_2 u_1 u_2\,\tfrac{\cos\epsilon}{r}. \quad\text{.......................}(9)$$

Comparing this with (8) we see that there is a notable difference.

7. The mutual energy being different, the forces on the charges, as derived by J. J. Thomson by the use of Lagrange's equations, will be different. When the speeds are constant, we shall have simply the

18

before-described vector product (4) for the "electromagnetic force;" or

$$\mathbf{F}_1 = \mu_0 q_1 \mathbf{V} u_1 \mathbf{H}_2, \qquad \mathbf{F}_2 = \mu_0 q_2 \mathbf{V} u_2 \mathbf{H}_1, \qquad \text{------------(10)}$$

if $\mathbf{F}_1$ is the electromagnetic force on the first and $\mathbf{F}_2$ that on the second element, whilst $\mathbf{H}_1$ and $\mathbf{H}_2$ are the magnetic forces. Similar changes are needed in the other parts of the complete mechanical forces.

It may be remarked that (if my calculations are correct) equation (7) or its equivalents expresses the mutual energy of any two rational current-elements (see § 1) in a medium of uniform inductivity, of moments $q_1 u_1$, and $q_2 u_2$, whether the currents be of displacement, or conduction, or convection, or all mixed, it being in fact the mutual energy of a pair of definite magnetic fields. But, since the hypothesis of instantaneous action is expressly involved in the above, the application of (7) is of a limited nature.

General Theory of Convection Currents.

8. Now leaving behind altogether the subject of current-elements, in the investigation of which one is liable to be led away from physical considerations and become involved in mere exercises in differential coefficients, and coming to the question of the electromagnetic effects of a charge moving in any way, I have been agreeably surprised to find that my solution in the case of steady rectilinear motion, originally an infinite series of corrections, easily reduces to a very simple and interesting finite form, provided u be not greater than v. Only when $u>v$ is there any difficulty. We must first settle upon what basis to work. First the Faraday-law (p standing for d/dt),

$$-\operatorname{curl}\mathbf{E}=\mu_0 p\mathbf{H}, \quad\cdots\cdots\cdots\cdots\cdots\cdots\cdots(11)$$

requires no change when there is moving electrification. But the analogous law of Maxwell, which I understand to be really a *definition* of electric current in terms of magnetic force, (or a doctrine), requires modification if the true current is to be

$$\mathbf{C}+p\mathbf{D}+\rho\mathbf{u}; \quad\cdots\cdots\cdots\cdots\cdots\cdots\cdots\cdots(12)$$

viz. the sum of conduction-current, displacement-current, and convection-current $\rho\mathbf{u}$, where ρ is the volume-density of electrification. The addition of the term $\rho\mathbf{u}$ was, I believe, proposed by G. F. Fitzgerald.[7]

(This was not meant exactly for a new proposal, being in fact after Rowland's experiments; besides which, Maxwell was well acquainted with the idea of a convection-current. But what is very strange is that Maxwell, who insisted so strongly upon his doctrine of the *quasi*-incompressibility of electricity, never formulated the convection-current in his treatise. Now Prof. Fitzgerald pointed out that if Maxwell, in his equation of mechanical force,

$$\mathbf{F}=\mathbf{VCB}-e\nabla\Psi-m\nabla\Omega,$$

had written $\mathbf{E}$ for $-\nabla\Psi$, as it is obvious he should have done, then the inclusion of convection-current in the true current would have followed naturally. (Here $\mathbf{C}$ is the true current, $\mathbf{B}$ the induction, e the density of electrification, m that of imaginary magnetic matter, Ψ the electrostatic and Ω the magnetic potential, and $\mathbf{E}$ the real electric force.)

Now to this remark I have to add that it is as unjustifiable to derive $\mathbf{H}$ from Ω as $\mathbf{E}$ from ψ; that is, in general, the magnetic force is not the slope of a scalar potential; so, for $-\nabla\Omega$ we should write $\mathbf{H}$, the real magnetic force.

But this is not all. There is possibly a fourth term in $\mathbf{F}$, expressed by $4\pi\mathbf{VDG}$, where $\mathbf{D}$ is the displacement and $\mathbf{G}$ the magnetic current; I have termed this force the "magneto-electric force," because it is the analogue of Maxwell's "electromagnetic force,"$\mathbf{VCB}$. Perhaps the simplest way of deriving it is from Maxwell's electric stress, which was the method I followed.[8]

Thus, in a homogeneous nonconducting dielectric free from electrification and magnetization, the mechanical force is the sum of the "electromagnetic" and the "magnetoelectric," and is given by

$$\mathbf{F}=\frac{1}{v^2}\frac{d\mathbf{W}}{dt},$$

where $\mathbf{W}=V\mathbf{EH}/4\pi$ is the transfer-of-energy vector.

It must, however, be confessed that the real distribution of the stresses, and therefore of the forces is open to question. And when ether is the medium, the mechanical force in it, as for instance in a light-wave, or in a wave sent along a telegraph-circuit, is not easily to be interpreted.)

The companion to (11) in a nonconducting dielectric is now

$$\mathrm{curl}\ \mathbf{H}=cp\mathbf{E}+4\pi\rho u. \qquad\text{------------------------}(13)$$

Eliminate $\mathbf{E}$ between (11) and (13), remembering that $\mathbf{H}$ is circuital, because μ_0 is constant, and we get

$$\left(p^2/v^2-\nabla^2\right)\mathbf{H}=\mathrm{curl}\ 4\pi\rho u, \qquad\text{----------------}(14)$$

the characteristic of $\mathbf{H}$. Here $\nabla^2=d^2/dx^2+\cdots$, as usual.

Comparing (14) with the characteristic of $\mathbf{H}$ when there is impressed force e instead of electrification ρ, which is

$$\left(p^2/v^2-\nabla^2\right)\mathbf{H}=\mathrm{curl}\ cpe,$$

we see that ρu becomes $cp\ e/4\pi$. We may therefore regard convection-current as *impressed* electric current. From this comparison also we may see that an infinite plane sheet of electrification of uniform density cannot produce magnetic force by motion perpendicular to its plane. Also we see that the sources of disturbances when ρ is moved are the places where ρu has curl; for example, a dielectric sphere uniformly filled with electrification (which is imaginable), when moved, starts the magnetic force solely upon its boundary.

The presence of "curl" on the right side tells us, as a matter of mathematical simplicity, to make H/curl the variable. Let

$$\mathbf{H}=\mathrm{curl}\ \mathbf{A}, \qquad\text{----------------------------------}(15)$$

and calculate $\mathbf{A}$, which may be any vector satisfying (15). Its characteristic is

$$\left(p^2/v^2-\nabla^2\right)\mathbf{A}=4\pi\rho u. \qquad\text{---------------------}(16)$$

The divergence of $\mathbf{A}$ is of no moment, and it is only vexatious complication to introduce ψ. The time-rate of decrease of $\mathbf{A}$ is not the real distribution of electric force, which has to be found by the additional datum

$$div\ c\mathbf{E}=4\pi\rho, \qquad\text{--}(17)$$

where $\mathbf{E}$ is the real force.

9. "Symbolically" expressed, the solution of (16) is

21

$$\mathbf{A} = \frac{4\pi\rho\mathbf{u}}{p^2/v^2 - \nabla^2} = \frac{-4\pi\rho\mathbf{u}/\nabla^2}{1 - p^2/v^2\nabla^2}. \quad\quad\quad\quad (18)$$

Here the numerator of the fraction to the right is the vector-potential of the convection-current. Calling it $\mathbf{A}_0$, we have

$$\mathbf{A}_0 = \frac{4\pi\rho\mathbf{u}}{-\nabla^2} = \sum \frac{\rho\mathbf{u}}{r}. \quad\quad\quad\quad (19)$$

Inserting in (18) and expanding, we have

$$\mathbf{A} = \left\{ 1 + (p/v\nabla)^2 + (p/v\nabla)^4 + \cdots \right\} \mathbf{A}_0. \quad\quad\quad (20)$$

Given then $\rho\mathbf{u}$ as a function of position and time, $\mathbf{A}_0$ is known by (19), and (20) finds $\mathbf{A}$, whilst (15) finds $\mathbf{H}$.

Complete Solution in the Case of Steady Rectilinear Motion. Physical Inanity of Ψ.

10. When the motion of the electrification is all in one direction, say parallel to the z-axis, u, A_0, and A are all parallel to this axis, so that we need only consider their tensors. When there is simply one charge q at a point, we have

$$A_0 = qu/r,$$

and (20) becomes

$$A = q\left\{1 + (p/v\nabla)^2 + (p/v\nabla)^4 + \cdots\right\}(u/r) \qquad \cdots\cdots\cdots\cdots(21)$$

at distance r from q. When the motion is steady, and the whole electromagnetic field is ultimately steady with respect to the moving charge, we shall have, taking it as origin,

$$p = -u(d/dz) = -uD$$

for brevity; so that

$$A = qu\left\{1 + (uD/v\nabla)^2 + (uD/v\nabla)^4 + \cdots\right\}r^{-1}. \qquad \cdots\cdots\cdots(22)$$

Now the property

$$\nabla^2 r^{n+2} = (n-2)(n+3)r^n \qquad \cdots\cdots\cdots\cdots\cdots\cdots(23)$$

brings (22) to

$$A = qu\left\{\frac{1}{r} + \frac{u^2}{v^2}D^2\frac{r}{2!} + \frac{u^4}{v^4}D^4\frac{r^3}{4!} + \cdots\right\}; \qquad \cdots\cdots\cdots\cdots(24)$$

and the property

$$D^{2n}r^{2n-1} = 1^2.3^2.5^2\cdots(2n-1)^2 v^{2n}/r, \qquad \cdots\cdots\cdots\cdots(25)$$

where $v = \sin\theta$, θ being the angle between $\mathbf{r}$ and the axis, brings (24) to

$$A = \frac{qu}{r}\left\{1 + \frac{u^2}{v^2}\frac{v^2}{2}(1 + \frac{u^2}{v^2}\frac{3}{4}v^2(1 + \frac{u^2}{v^2}\frac{5}{6}v^2(1 + \cdots\right\}; \qquad \cdots\cdots(26)$$

which, by the Binomial Theorem, is the same as

$$A = (qu/r)\left\{1 - u^2 v^2/v^2\right\}^{-\frac{1}{2}}, \qquad \cdots\cdots\cdots\cdots\cdots\cdots(27)$$

the required solution.

11. To derive H, the tensor of the circular $\mathbf{H}$, let $rv = h$, the distance from the axis. Then, by (15),

$$H = -\frac{dA}{dh} = -\nu\frac{dA}{dr} + \frac{\mu\nu}{r}\frac{dA}{d\mu} = \frac{qu\nu}{r^2}\left(1 + \mu\frac{d}{d\mu}\right)\left(1 - \frac{u^2}{v^2}v^2\right)^{-\frac{1}{2}}, \qquad (28)$$

by (27), if $\mu = \cos\theta$. Performing the differentiation, and also getting out E the tensor of the electric force, we have the final result that the electromagnetic field is fully given by[9]

$$cE = \frac{q}{r^2} \cdot \frac{1-u^2/v^2}{(1-u^2v^2/v^2)^{\frac{3}{2}}}, \qquad H = cEuv, \qquad \cdots\cdots\cdots\cdots\cdots\cdots(29)$$

with the additional information that **E** is radial and **H** circular.

Now, as regards Ψ, if we bring it in, we have only got to take it out again. When the speed is very slow we may regard the electric field as given by $-\nabla\Psi$ *plus* a small correcting vector, which we may call the electric force of inertia. But to show the *physical* inanity of Ψ, go to the other extreme, and let u nearly equal v. It is now the electric force of inertia (supposed) that equals $+\nabla\Psi$ nearly (except about the equatorial plane), and its sole utility or function is to cancel the other $-\nabla\Psi$ of the (supposed) electrostatic field. It is surely impossible to attach any physical meaning to Ψ and to propagate it, for we require two Ψ's, one to cancel the other, and both propagated infinitely rapidly.

As the speed increases, the electromagnetic field concentrates itself more and more about the equatorial plane, $\theta = \frac{1}{2}\pi$. To give an idea of the accumulation, let $u^2/v^2 = .99$. Then cE is .01 of the normal value q/r^2 at the pole, and 10 times the normal value at the equator. The latitude where the value is normal is given by

$$\nu = (v/u)\left[1-(1-u^2/v^2)^{\frac{2}{3}}\right]^{\frac{1}{2}}. \qquad \cdots\cdots\cdots\cdots\cdots\cdots\cdots\cdots\cdots\cdots\cdots(30)$$

Limiting Case of Motion at the Speed of Light. Applications to a Telegraph Circuit.

12. When $u=v$, the solution (29) becomes a plane electromagnetic wave, E and H being zero everywhere except in the equatorial plane. As, however, the values of E and H are infinite, distribute the charge along a straight line moving in its own line and let the linear-density be q. The solution is then[10]

$$H=Ecv=2qv/r \quad \cdots\cdots\cdots\cdots\cdots\cdots\cdots\cdots\cdots\cdots\cdots\cdots\cdots\cdots(31)$$

at distance r from the line, between the two planes through the ends of the line perpendicular to it, and zero elsewhere.

To further realize, let the field terminate internally at $r=a$, giving a cylindrical-surface distribution of electrification, and terminate the tubes of displacement externally upon a coaxial cylindrical surface; we then produce a real electromagnetic plane wave with electrification, and of finite energy. We have supposed the electrification to be carried through the dielectric at speed v, to keep up with the wave, which would of course break up if the charge were stopped. But if perfectly conducting surfaces be given on which to terminate the displacement, the natural motion of the wave will itself carry the electrification along them. In fact we now have the rudimentary telegraph-circuit, with no allowance made for absorption of energy in the wires, and the consequent distortion. If the conductors be not coaxial, we only alter the distribution of the displacement and induction, without affecting the propagation without distortion.[11]

If we now make the medium conduct electrically, and likewise magnetically, with equal rates of subsidence, we shall have the same solutions, with a time-factor $\varepsilon^{-\rho t}$ producing ultimate subsidence to zero; and, with only the real electric conductivity in the medium the wave is running through, it will approximately cancel the distortion produced by the resistance of the wires the wave is passing over when this resistance has a certain value.[12] We should notice, however, that it could not do so perfectly, even if the magnetic retardation in the wires due to diffusion were zero; because in the case of the unreal magnetic conductivity its correcting influence is where it is wanted to be, in the body of the wave; whereas in the case of the wires, their resistance, correcting the distortion due to the external conductivity, is outside the wave; so that we virtually assume instantaneous propagation laterally from the wires of their correcting influence in the elementary theory of propagation along a telegraph-circuit which is symbolized by the equations

$$-dV/dz=(R+Lp)C, \qquad -dC/dz=(K+Sp)V \quad \cdots\cdots\cdots(32)$$

where R, L, K, and S are the resistance, inductance, leakage-conductance, and permittance per unit length of circuit, C the current, and V what I, for convenience, term the potential-difference, but which I have expressly disclaimed[13] to represent the electrostatic difference of potential, and have shown to represent the transverse voltage or line-integral of the electric force across the circuit from wire to wire,

25

including the electric force of inertia. Now in case of great distortion, as in a long submarine cable, this V approximates towards the electrostatic potential-difference, which it is in Sir W. Thomson's diffusion theory; but in case of little distortion, as in telephony through circuits of low resistance and large inductance, there may be a wide difference between my V and that of the electrostatic force. Consider, for instance, the extreme case of an isolated plane-wave disturbance with no spreading-out of the tubes of displacement. At the boundaries of the disturbance the difference between V and the electrostatic difference of potential is great.

But it is worth noticing, as a rather remarkable circumstance, that when we derive the system (32) by elementary considerations, viz. by extending the diffusion-system by the addition of the E.M.F. of inertia and leakage-current, we apparently as a matter of course take V to mean the same as in the diffusion-system. The resulting equations are correct, and yet the assumption is certainly wrong. The true way appears to be that given by me in the paper last referred to, by considering the line-integral of electric force in a closed curve. [vol. II., p. 187. Also p. 87]. We cannot, indeed, make a separation of the electric force of inertia from $-\nabla\Psi$ without some assumption, though the former is quite definite when the latter is suitably defined. But, and this is the really important matter, it would be in the highest degree inconvenient, and lead to much complication and some confusion, to split V into two components, in other words, to bring in Ψ and $\mathbf{A}$.

In thus running down Ψ, I am by no means forgetful of its utility in other cases. But it has perhaps been greatly misused. The clearest course to pursue appears to me to invariably make $\mathbf{E}$ and $\mathbf{H}$ the primary objects of attention, and only use potentials when they naturally suggest themselves as labour-saving appliances.

Special Tests. The Connecting Equations.

13. Returning to the solutions (29), the following are the special tests of their accuracy. Let E_1 and E_2 be the z and h components of **E**. Then by (11) and (13), with the special meaning assumed by p, we have

$$\left. \begin{array}{rclcl} \frac{1}{h}\frac{d}{dh}hH & = & -cu\frac{dE_1}{dz}, & & \\[4pt] -\frac{dH}{dz} &=& -cu\frac{dE_2}{dz}, & \text{or} & H=cuE_2 \\[4pt] \frac{dE_1}{dh}-\frac{dE_2}{dz} &=& -\mu_0 u\frac{dH}{dz}, & \text{or} & \frac{dE_1}{dh}=\left(1-\frac{u^2}{v^2}\right)\frac{dE_2}{dz}. \end{array} \right\} \quad \cdots(33)$$

In addition to satisfying these equations, the displacement outward through any spherical surface centred at the charge may be verified to be q; this completes the test of the accuracy of (29).

But (33) are not limited to the case of a single point-charge, being true outside the electrification when there is symmetry with respect to the z-axis, and the electrification is a l moving parallel to it at speed u.

When $u=v$, $E_1=0$, and $E_2=E=\mu v H$, so that we reduce to

$$\frac{1}{h}\frac{d}{dh}hH=0, \quad \text{---} (34)$$

outside the electrification. Thus, if the electrification is on the axis of z, we have

$$E/\mu v = H = 2qv/r, \quad \text{--} (35)$$

differing from (31) only in that q, the linear density, may be any function of z.

The Motion of a Charged Sphere. The Condition at a Surface of Equilibrium (Footnote).

14. If, in the solutions (29), we terminate the fields internally at $r=a$, the perpendicularity of **E** and the tangentiality of **H** to the surface show that (29) represents the solutions in the case of a perfectly conducting sphere of radius a, moving steadily along the z-axis at the speed u, and possessing a total charge q. The energy is now finite. Let U be the total electric and T the total magnetic energy. By space-integration of the squares of E and H we find that they are given by

$$U = \frac{q^2}{2ca} \cdot \frac{1-u^2/v^2}{4} \left[1 + \frac{\frac{3}{2}}{1-u^2/v^2} + \frac{\frac{3}{2}\tan^{-1}\frac{u/v}{(1-u^2/v^2)^{\frac{1}{2}}}}{(u/v)(1-u^2/v^2)^{\frac{1}{2}}} \right], \quad \cdots\cdots\cdots\cdots(36)$$

$$T = \frac{q^2}{2ca} \cdot \frac{1-u^2/v^2}{4} \left[1 + \frac{2u^2/v^2 - \frac{1}{2}}{1-u^2/v^2} + \frac{\left(2u^2/v^2 - \frac{1}{2}\right)\tan^{-1}\frac{u/v}{(1-u^2/v^2)^{\frac{1}{2}}}}{(u/v)(1-u^2/v^2)^{\frac{3}{2}}} \right], \quad (37)$$

in which $u<v$. When $u=v$, with accumulation of the charge at the equator of the sphere, we have infinite values, and it appears to be only possible to have finite values by making a zone at the equator cylindrical instead of spherical. The expression for T in (37) looks quite wrong; but it correctly reduces to that of equation (2) when u/v is infinitely small.[14]

28

The State when the Speed of Light is exceeded.

15. The question now suggests itself, What is the state of things when $u>v$? It is clear, in the first place, that there can be no disturbance at all in front of the moving charge (at a point, for simplicity). Next, considering that the spherical waves emitted by the charge in its motion along the z-axis travel at speed v, the locus of their fronts is a conical surface whose apex is at the charge itself, whose axis is that of z, and whose semiangle θ is given by

$$\sin \theta = v/u. \quad \cdots\cdots\cdots\cdots\cdots\cdots\cdots\cdots(38)$$

The whole displacement of amount q, should therefore lie within this cone. And since the moving charge is a convection-current qu, the displacement-current should be towards the apex in the axial portion of the cone, and change sign at some unknown distance, so as to be away from the apex either in the outer part of the cone or else upon its boundary. The pulling back of the charge by the electric stress would require the continued application of impressed force to keep up the motion, and its activity would be accounted for by the continuous addition made to the energy in the cone; for the transfer of energy on its boundary is perpendicularly outward, and the field at the apex is being continuously renewed.

The above general reasoning seems plausible enough, but I cannot find any solution to correspond that will satisfy all the necessary conditions. It is clear that (29) will not do when $u>v$. Nor is it of any use to change the sign of the quantity under the radical, when needed, to make real. It is suggested that whilst there should be a definite solution, there cannot be one representing a *steady* condition of **E** and **H** with respect to the moving charge. As regards physical possibility, in connexion with the structure of the ether, that is not in question.[15]

A Charged Straight Line moving in its own Line.

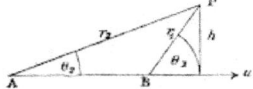

16. Let us now derive from (29), or from (27), the results in some cases of distributed electrification, in steady rectilinear motion. The integrations to be effected being all of an elementary character, it is not necessary to give the working.

First, let a straight line AB be charged to linear density q, and be in motion at speed u in its own line from left to right. Then at P we shall have

$$A = qu \ \log\left(\frac{r_1}{r_2} \cdot \frac{\mu_1 + \left(1 - v_1^2 u^2 / v^2\right)^{\frac{1}{2}}}{\mu_2 + \left(1 - v_2^2 u^2 / v^2\right)^{\frac{1}{2}}} \right), \quad \text{...........................(39)}$$

from which $H = -dA/dh$ gives

$$H = qu\left(1 - \frac{u^2}{v^2}\right)\left[\frac{v_1}{r_1\left(1 - v_1^2 u^2 / v^2\right) + r_1 \mu_1 \left(1 - v_1^2 u^2 / v^2\right)^{\frac{1}{2}}} - \text{same } f^n \text{ of } r_2, \mu_2, v_2 \right], \quad (40)$$

where $\mu = \cos \theta$, $v = \sin \theta$.

When P is vertically over B, and A is at an infinite distance, we shall find

$$H = qu/h, \quad \text{...(41)}$$

which is one half the value due to an infinitely long (both ways) straight current of strength qu. The notable thing is the independence of the ratio u/v.

But if $u=v$ in (40), the result is zero, unless $v_1=1$, when we have the result (41). But if P be still further to the left, we shall have to add to (41) the solution due to the electrification which is ahead of P. So when the line is infinitely long both ways, we have double the result in (41), with independence of u/v again. But should q be a function of z, we do not have independence of u/v except in the already considered case of $u=v$, with plane waves, and no component of electric force parallel to the line of motion.

A Charged Straight Line moving Transversely.

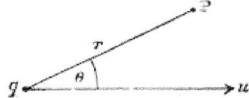

17. Next, let the electrified line be in steady motion perpendicularly to its length. Let q be the linear density (constant), the z-axis that of the motion, the x-axis coincident with the electrified line and that of y upward on the paper. Then the A at P will be

$$A=\frac{qu}{(1-u^2/v^2)} \log \frac{x_1+\left\{x_1^2+y^2+z^2(1-u^2/v^2)-1\right\}^{\frac{1}{2}}}{x_2+\left\{x_2^2+y^2+z^2(1-u^2/v^2)-1\right\}^{\frac{1}{2}}} \qquad \cdots\cdots\cdots(42)$$

where y and z belong to P, and x_1, x_2 are the limiting values of x in the charged line. From this derive the solution in the case of an infinitely long line. It is

$$cE=\frac{2q}{r}\cdot\frac{(1-u^2/v^2)^{\frac{1}{2}}}{1-v^2u^2/v^2}, \qquad H=cEuv, \qquad \cdots\cdots \cdots\cdots(43)$$

where v=sin θ understanding that **E** is radial, or along q P in the figure, and **H** rectilinear, parallel to the charged line. Terminating the fields internally at $r=a$, we have the case of a perfectly conducting cylinder of radius a, charged with q per unit of length, moving transversely. When $u=v$ there is disappearance of E and H everywhere except in the plane $\theta=\frac{1}{2}\pi$, as in the case of the sphere, and consequent infinite values. It is the curvature that permits this to occur, i.e. producing infinite values; of course it is the self-induction that is the cause of the conversion to a plane wave, here and in the other cases. There is some similarity between (43) and (29). In fact, (43) is the bidimensional equivalent of (29).

31

A Charged Plane moving Transversely.

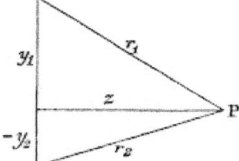

18. Coming next to a plane distribution of electrification, let q be the surface-density, and the plane be moving perpendicularly to itself. Let it be of finite breadth and of infinite length, so that we may calculate H from (43). The result at P is

$$H = \frac{qu}{(1-u^2/v^2)^{\frac{1}{2}}} \log \frac{r_1^2 - y_1^2 u^2/v^2}{r_2^2 - y_2^2 u^2/v^2}. \quad \cdots\cdots\cdots\cdots\cdots\cdots(44)$$

When P is equidistant from the edges, H is zero. There is therefore no H anywhere due to the motion of an infinitely large uniformly charged plane perpendicularly to itself. The displacement-current is the negative of the convection-current and at the same place, viz. the moving plane, so there is no true current.

Calculating E_1; the z-component of $\mathbf{E}$, z being measured from left to right, we find

$$cE_1 = 2q \left\{ \tan^{-1} \frac{y_1}{z} \left(1 - \frac{u^2}{v^2} \right)^{\frac{1}{2}} - \tan^{-1} \frac{y_2}{z} \left(1 - \frac{u^2}{v^2} \right)^{\frac{1}{2}} \right\} \quad (45)$$

The component parallel to the plane is H/cu. Thus, when the plane is infinite, this component vanishes with H, and we are left with

$$cE_1 = cE = 2\pi q, \quad (46)$$

the same as if the plane were at rest.

32

A Charged Plane moving in its own Plane.

19. Lastly, let the charged plane be moving in its own plane. Refer to the first figure, in which let AB now be the trace of the plane when of finite breadth. We shall find that

$$H = 2qu \left[\tan^{-1} \frac{z}{h(1 - u^2/v^2)^{\frac{1}{2}}} \right]_{z_1,}^{z_2} \tag{47}$$

z_1 and z_2 being the extreme values of z, which is measured parallel to the breadth of the plane.

Therefore, when the plane extends infinitely both ways, we have

$$H = 2\pi qu \tag{48}$$

above the plane, and its negative below it. This differs from the previous case of vanishing displacement-current. There is H, and the convection-current is not now cancelled by coexistent displacement-current.

The existence of displacement-current, or changing displacement, was the basis of the conclusion that moving electrification constitutes a part of the true current. Now in the problem (48) the displacement-current has gone, so that the existence of H appears to rest merely upon the assumption that moving electrification is true current. But if the plane be not infinite, though large, we shall have (48) nearly true near it, and away from the edges whilst the displacement-current will be strong near the edges and almost nil where (48) is nearly true.

But in some cases of rotating electrification, there need be no displacement anywhere, except during the setting up of the final state. This brings us to the rather curious question whether there is any difference between the magnetic field of a convection-current produced by the rotation of electrification upon a good nonconductor and upon a good conductor respectively, other than that due to diffusion in the conductor. For in the case of a perfect conductor, it is easy to imagine that the electrification could be at rest, and the moved conductor merely slip past it. Perhaps Professor Rowland's forthcoming experiments on convection-currents may cast some light upon this matter.

December 27, 1888.

---

1. *Phil. Mag.*, April, 1881.

2. "Applications of Dynamics to Physics and Chemistry," chap. iv. pp. 31 to 37.

3. *The Electrician*, Jan. 24, 1885, p. 220. [vol. I., p. 446].

4. *The Electrician*, Dec. 28, 1888, p. 230. [p. 501, vol. II.].

5. *The Electrician*, Jan 24, 1885, p. 221. [vol.I., p. 446].

6. "Applications of Dynamics to Physics and Chemistry," chap. iv.; and *Phil. Mag.*, April 1881.

7. Brit. Assoc., Southport, 1883.

8. "El. Mag. Ind. and its Prop." XXII. *The Electrician,* Jan. 15, 1886, p. 187. [vol. I., p. 545].

9. *The Electrician,* Dec. 7, 1888, p. 148 [p. 495, vol. II].

10. Ibid. Nov. 23, 1888, p. 84. [p. 493, vol. II.].

11. *The Electrician,* Jan. 10, 1885. [p. 440, vol. I.]. Also "Self-Induction of Wires." Part IV. *Phil. Mag.*, Nov. 1886. [p. 221, vol. II.].

12. "Electromagnetic Waves," § 6, *Phil. Mag.* Feb. 1888 [p. 379, vol. II.]. *The Electrician*, June 1887 [p. 123, vol. II.].

13. "Self-Induction of Wires," part II. *Phil. Mag.* Sept 1886 [vol. II. p. 189].

14. [I am indebted to Mr. G. F. C. Searle, of Cambridge, for the opportunity of making a somewhat important correction before going to press. In a private communication (August 19, 1892) he informed me that he had verified the accuracy of the solution for a point-charge, which he had also obtained in another way, from equations equivalent to (33), without the use of the function **A** of §§ 8 to 10; but he cast doubt upon the validity of the extension made in § 14, from a point-charge to a charged conducting sphere, and asked the plain question (in effect), What justification is there for terminating the displacement perpendicularly, to make a surface of equilibrium?

On examination, I find that there is no justification whatever, exceptions excepted. The true boundary condition may, however, be found without a fresh investigation. On p. 499 the problem of uniform motion of electrification through a dielectric medium, or conversely, of the uniform motion of the whole medium past stationary electrification, is reduced to a case of eolotropy in electrostatics. The effect of the motion of the isotropic medium on the displacement emanating from stationary electrification is there shown to be identical with the effect of keeping the medium stationary and reducing its permittivity in lines parallel to the (abolished) motion from c to $c(1-u^2/v^2)$, whilst keeping the transverse permittivity the same. The transverse concentration of the displacement is obvious. Now the function P (equation (14), p. 499) is the electrostatic potential in the stationary eolotropic problem, so that its slope $-\nabla P$, which call **F**, is the electric force, and the displacement **D** is a linear function thereof, say **D**=λ**F**, where λ is the permittivity operator. The condition of equilibrium is that **F** is perpendicular to the surface where it terminates, this being required to make curl **F**=0, or the voltage zero in every circuit. Now, in the corresponding problem of the same electrification in a moving isotropic medium, we have the same function P (no longer the electrostatic potential) and the same derived vector **F**, whilst the displacement **D** is also derived from **F** in the same way. But whilst the meaning of **D** is the same in both cases, that of **F** is not. In the eolotropic case, **F** is the electric force, and is not parallel to **D**. In the moving isotropic medium, on the other hand, **F** is not the electric force, which is **E**, parallel to **D**. Nevertheless, the same condition formally obtains, for we have {{{1}}} in the moving medium, requiring that **F** shall be perpendicular to a surface of equilibrium, not the electric force or displacement. P=constant is therefore the equation to a surface of equilibrium. That is, in the case of a point-charge, the surfaces of equilibrium are not spheres, but are concentric oblate spheroids, whose principal axes are proportional to the square roots of c, c, and $c(1-u^2/v^2)$, the principal permittivities in the eolotropic problem. In the extreme case of $u=v$, the spheroid reduces to a flat circular disc, with a single circular line of electrification on its edge. It would seem, however, to be a matter of indifference, in this extreme case, whether the conductor be a disc or a solid sphere. Bearing in mind the conditions assumed to prevail in the problem of motion of sources of displacement in a uniform medium, we see that if we introduce conductors, say by filling up spaces void of electric force with conducting matter, this should not interfere with the assumed motions. (See also "Electromagnetic Theory," § 164.)

Equations (36) (37) express the electric and magnetic energy outside a sphere of radius a, within which is either a point-source at the origin, or any equivalent spheroidal electrified surface.

In the corresponding bidimensional problem of § 17 in the text, with the solution (43), it is clear from the above that the surface of equilibrium is an elliptic cylinder, the shorter axis being in the direction of motion, and the axes themselves in the ratio 1 to $\left(1-u^2/v^2\right)^{\frac{1}{2}}$. This surface degenerates to a flat strip when $u=v$.]

15. [The difficulty about the above method and solution (29) is that it is not explicit enough when $u>v$, and does not indicate the limits of application. It gives a real solution for the hinder cone, a real solution for the forward cone, and an unreal solution in the rest of space, but we have no instruction to reject the part for the forward cone and the unreal part, nor have we any means of testing that the remainder, confined to the hinder cone, is the proper solution, viz., by the test of divergence, to give the right amount of electrification. The integral displacement comes to $-\infty$. Now this may require to be supplemented by $+\infty+q$ on the boundary of the cone, but we have no way of testing   it.

But certain considerations led me to the conclusion that the problem of $u>v$ was really quite as definite a one as that of $u<v$, and that a correct method of a general character (independent of the magnitude of u) would show this explicitly. I therefore (in 1890) attacked the problem from a different point of view, employing the method of resistance-operators (or an equivalent method). Form the complete differential equation $\mathbf{D}=\varphi\mathbf{u}$, connecting the displacement $\mathbf{D}$ associated with a moving point-charge with its velocity u, which is any function of the time t. Here φ is a differential operator, a function of p or d/dt. The solution of this equation gives $\mathbf{D}$ explicitly in terms of $\mathbf{u}$, whether steady or variable, and its structure indicates the limits of application.

Taking $\mathbf{u}$=constant, we obtain the result (29) when $u<v$. But when $u>v$, the formula tells us to exclude all space except the hinder cone, and that in it, the solution is *not* (29), but *double as much*. That is, double the right member of the first of (29) when $u>v$. The boundary of the cone is also a displacement sheet. The displacement is to the charge in the cone, and from the charge on its surface. Being so near the end of the second volume, I regret that there is no space here for the mathematical investigation, which cannot be given in a few words, and must be reserved.]

35

www.ingramcontent.com/pod-product-compliance
Lightning Source LLC
Chambersburg PA
CBHW070746180526
45168CB00004B/1552